Mapping the Future: Rising Stars in GIS

—

Georgia Mackay

CONTENTS

ACKNOWLEDGMENTS

Without the following, my journey and this book
would simply not be possible:

Kym Kelly & Ron Holmes –
My first two mentors/supervisors in GIS.
For taking a chance on me. For your substantial positive
impact on the start of my GIS career.

Bill Shields –
For strength, mentorship, faith, and always reminding me
that there are many ways to skin a cat. For helping me
with more than I can list. I hope to be like you someday.

To those I interviewed –
I am happy and proud to be a part of your journey.

INTRODUCTION

A Geographic Information System (GIS) is a powerful technology that allows us to interact with and map the world around us. It consists of XY coordinate data and its associated attributes. To put it plainly, if you can point to it on a map and say something about it, you have the beginnings of a GIS database. GIS users leverage maps and data to provide insights into geographical patterns. By visualizing spatial data, we can facilitate informed decision-making across various fields including (but not limited to) urban planning, environment management, and public health.

This book explores the lives of awesome newcomers in the GIS world. By reading their stories, you will see the future of the industry. The narratives in this book were captured in a series of interviews and paraphrased to allow for more structured reading. On the following page are the questions I asked the interviewees to prompt relevant information about their past, present, and future in GIS. Interviewees were provided with the questions in advance.

MAPPING THE FUTURE: RISING STARS IN GIS

1. Can you briefly introduce yourself and provide some insight into your educational background, especially any coursework or projects related to GIS?

2. **What was your early understanding of GIS?** What motivated you to pursue a career in the GIS industry, and how do you envision your educational background benefiting your work in GIS?

3. Can you share any hobbies or interests that you believe have contributed to your personal and professional growth, even if they aren't directly related to GIS?

4. Are there specific qualities or character traits that you consider essential for success in the GIS field? How do you demonstrate these traits in your work?

5. While you may have limited work experience, could you describe any GIS-related projects or activities you've been involved in, either during your studies or early in your career?

6. What aspects of GIS are you most passionate about, and how have you worked to develop your skills and knowledge in these areas?

7. What are your career goals and aspirations in the GIS field, and what steps are you currently taking to achieve them?

8. Have there been any GIS professionals, books, or resources that have inspired or influenced your understanding of GIS, even at this early stage in your career?

9. In your opinion, what are the most exciting possibilities or trends in GIS that you've encountered so far, and how do you foresee these developments shaping the future of the industry?

10. How do you plan to continue learning and stay updated with the latest developments in GIS? **What does GIS mean to you now?**

Why these questions? I felt like they established the right amount of personal identity and qualifications related to GIS.

The first three questions introduce each person. Understanding what motivated the interviewee to pursue a career in GIS sheds light on their passion and commitment to the industry. Exploring how their education benefits their work reveals the relevance of their academic background. Hobbies and interests can offer insights into their well-roundedness and potential for creativity.

Asking about the qualities or character traits they deem essential for GIS success reveals the interviewee's self-awareness and understanding of the industry's demands. Demonstrating these traits in their work showcases their practical application. Even with limited work experience, Question 5 encourages the interviewee to share any hands-on GIS involvement, whether through coursework, internships, or early career projects. It demonstrates their practical exposure to the field. Then, we identify specific aspects of GIS that excite the interviewee and show where their interests lie. Discussing how they've developed their skills highlights their commitment to growth and specialization.

We then transition to each interviewee's career goals and the actions they're taking to achieve them. We acknowledge external influences on their GIS knowledge and their understanding. By asking about possibilities or trends in GIS, we explore their insights into exciting GIS possibilities and trends with a focus on the future. The final questions discuss their strategies for ongoing learning and professional development in the GIS field.

All interviews were concluded with an open prompt for the interviewee to share anything they felt the questions missed and if they would like to leave advice for other people starting their careers in GIS.

The two bold questions provide a natural beginning and end to each narrative.

GEORGIA MACKAY

I started my journey at Delaware Technical Community College. I sat in the career office with my father and a career counselor, and I simply had no idea what I wanted to go to school for.

On the one hand was my love of science. My father received MS and PhD degrees in Chemical Engineering and is now an internationally-known leader in nanotechnology. I participated in a program called FIRST Robotics in high school and have always had a great passion for STEM (Science, Technology, Engineering, Mathematics), but I knew I didn't want to go to school explicitly for computer science or robotics. I never envisioned myself as a programmer who writes code from 9am-5pm; I'm just someone who can code and use it occasionally as a tool.

On the other hand was my love of art, design, and graphical work. My mother is an interior designer with amazing spatial awareness, CAD (Computer Aided Design) skills, and is a wizard with a tape measure. I've always been involved in art; I'm a multi-instrumentalist and graphic design is truly a passion of mine.

After I spilled all this to the career counselor, she sat there, nodded, and said she had just the thing: A new emerging technology that is in high demand - GIS. I was instantly sold on studying the intersection of data analysis, programming, and cartography. My advisor at DTCC was also the mind behind the GIS program. In addition to all the basics of the GIS world, she taught me the art of planning and sticking to it, as well as how to look at problems from all angles. She picked adjunct instructors who worked in the "real world" outside of academia, so everyone who taught me there had real-world experience directly related to in-class materials. It created an amazing classroom experience that I feel really gave me a step up in the industry.

In my free time, I'm always up to something! If you ask my partner, he will say I am "bad at relaxing," but I would say I am just critically efficient with my time. Even when playing video games, the first thing I flesh out is the map. I keep extensive spreadsheets of all the collectables in each video game and where I found them (funnily enough, this itself is a GIS system). I feel that data management and metadata work is essential to finding success in GIS.

I have held four internships in GIS throughout my college experience and two professional career positions in GIS. I have worked with a local county to georeference over 500 aerial images from 1982 into a publicly available basemap. The image quality in these images is so high you can see the makes and models of cars in the mall parking lot. My very first supervisor was an amazing GIS Coordinator who took the time to sit with me and answer all my questions. He encouraged me to document my workflow and organized my very first public GIS presentation on the basemap to a local GIS user group. I still write my workflows in the format he taught me. After my time there, I went on to intern for the local county police department, where I created a private- and public-facing crime analysis map. Working on a GIS project that held police accountable during the height of the Black Lives Matter movement was an amazing experience. I've also extracted data from historic census documents, as well as managed a team to transcribe relevant information from historic work orders to catalog the locations of led pipes in a city. In my "big girl" jobs, I've worked as a GIS Analyst extracting features and terrain data, as well as mapping indoor spaces with a GIS toolset.

In GIS, I find that I am most passionate about the programming aspect. As much as I truly love picking final colors for a map component and what should be a serif or sans serif font, I often pause to think about any time-saving code I can write. If I am presented with clicking the same four buttons in the same order every hour, I would rather take the time to write a script to do it for me. Although the cost-benefit analysis of time spent may prove to rule in favor of repetitive action, my brain power can be better spent doing something non-repetitive.

My aspiration in the formal GIS field begins with writing a book about GIS. If you're reading this (and you're not my editor), I can cross that one off my list. I would also like to work towards my GISP (GIS Professional) title, a recognized and certified credential in the GIS world. You need various points in separate categories to apply to take the GISP test. I have enough points in Education and Work Experience but need more points in the Contributions category. I am also seeking to volunteer more, join user groups, and give presentations whenever I can.

I think the future of GIS will lead us towards more automation. AI that can navigate and make decisions based on the maps we create will be more commonplace: Smart drones that wayfind using maps, robots that navigate to the right hospital room, AI tour guides for campuses, and seamless GIS experiences for every walking trail and national park. Indoor GIS will be the key to breaking down the outside/inside barrier and I believe Indoor GIS is the next hot technology in the GIS sector.

I will be a lifelong learner. I believe that continuing to grow and expanding your knowledge will always lead you on the right path. When it comes to GIS, going to conferences and seeing other people's projects will help you broaden your horizons and inspire new possibilities. GIS means the world to me. Literally. It is how I view and understand the world around me;

it is a passion that drives me to explore, innovate, and make a difference. As I continue this exciting journey, I am excited to see where the ever-evolving field of GIS will take me, and how I can contribute to its ongoing transformation.

My advice to all others starting in GIS is as follows:

Apply yourself. Say 'yes' to every opportunity you are presented with. Be a go-getter and chase your passion. Never be afraid to ask questions! Always remember that it is a privilege to work with those with experience and those who know more than you. There's a saying: "If you're the smartest person in the room, you're in the wrong room." I greatly feel like that applies to GIS.

WILL CAMPBELL

Will began his journey in GIS at University of Wisconsin-Parkside, where he received a Bachelor's of Science in Environmental Science with minors in Biology and GIS. He went on to receive a Master's of Science from the University of Wisconsin-Madison in Cartography and GIS Development. Here, he learned how to make interactive map websites and applications while also digging into GIS Analysis.

He was introduced to GIS by his advisor at University of Wisconsin-Parkside. While working towards his degree in Environmental Science, his advisor suggested supplementing his education with an Introduction to GIS course. Will was hesitant at first; however, he soon discovered that GIS includes artistic expression through cartography, ultimately finding that "GIS is not only a type of science, but an art, as well." Will states that "it really depends on the cartographer to say what a 'good map' looks like," emphasizing the capacity for creativity and individuality in GIS.

After the Introduction course, he liked GIS so much that he pursued a minor in GIS and went on to get a Master's degree in it, feeling "that [his] education will benefit [his] work in GIS because not only [does he] understand what GIS is, but also the endless possibilities that making a GIS website has to offer!"

Will originally sought to follow his passion in Environmental Science, but his love for the environment and his passion for climate change solutions inspired him to find a job that mixes both fields. Will shares that his "GIS professors and supervisors have been a big influence in [his] career in GIS. [He] think[s] that almost everyone that [he's] met that is in GIS has an understanding that GIS doesn't always go to plan, and that's one of the big things that they helped [him] to understand."

His inaugural GIS project took place with NASA DEVELOP in Laramie, Wyoming, and was a significant milestone. The project involved analyzing imagery to generate a turbidity concentration map of a local river— a task demanding meticulous data analysis and the utilization of Python scripts and raster math. His work culminated in a thematic map delineating turbidity concentration level, shedding light on areas of high turbidity. "Turbidity is the clarity of a water body, and the more turbid a body of water is, the more sunlight it reflects. It generally comes from dirt or other pollutants." Although the team eventually followed a different avenue, Will still worked to create a thematic map of turbidity concentration in the river. In his end product, "you can then look at the data on the map and see different splotches on the map where the most turbidity is."

Will enjoys the feeling of analyzing and drawing GIS data with a strong focus on the cartographic aspect. He feels that one of the key traits to being successful in GIS is patience, "not only for satellite images that are taking forever to load onto your map, but also patience for others." On his previous project, he always made sure to get feedback from his team and ultimately "learned that people might have differing opinions on where the north arrow should go, or what color scheme looks the best, but as long as the map accurately conveys the message you are trying to send, that is what GIS is all about."

The next step on his path will be a GIS Technician or Analyst role, but he has dreams of coordinating projects or being a part of a management team for GIS work. He is hoping to work on a team that enables him to use data related to environmental science, further combining his passions. Will wants to work in ESRI's ArcGIS Pro; like many in the industry, he feels that it's the future of the GIS industry. He intends to continue his pursuit of GIS positions, connect with fellow GIS enthusiasts, and possibly engage in volunteer GIS work. Additionally, he aspires to obtain a GISP certification in the future.

To Will, "GIS is a lot more than 'basically Google Maps' like we tell most people. It is a tool that can be used for either good or evil, but it's up to the cartographer to follow the GIS Code of Conduct and hopefully use it for a good purpose instead of bending the data to look like something that it's not." He admits that in the beginning, learning the technical terminology of GIS can be confusing, but with practice, he feels that anyone can do GIS.

His advice to all others starting in GIS is as follows:

"No matter how much you feel like you might not belong in a field that you are passionate about, don't listen to that voice in your head or anyone that says you aren't enough because you are and you're awesome at what you do."

RENEE HUPP

Renee currently works as an Emergency Management Professional for a government agency. They hold several degrees: an Associate's in GIS, a Bachelor's in Homeland Security and Emergency Preparedness, and a Master's in Safety, Security, and Emergency Management. Though they have never specifically held a GIS position, they have worked on several projects related to GIS that they have "accidentally fallen into," such as radiological tracking and GIS, hurricane evacuation / population / storm surge data in GIS, and Covid and GIS.

Their earliest understanding of GIS was that it was a digital map to which people could contribute. They grew up with their dad, a bus driver and truck driver, reading "old style" paper maps. Renee states, "I already had an innate ability with maps and suddenly they were on the computer, and I can make them! I don't have to wait for a new edition to come out, I can start doing it. That was really cool to me." They got more involved with GIS by working with professionals in the state of Delaware, sharing that they enjoyed "the whole process," from research to crafting solutions that catered not only to visual learners, but also to those requiring rapid data analysis.

Renee and I have a shared hobby of Dungeons & Dragons, so we discussed how Renee makes their own maps for tabletop roleplaying games. We reflected on how the data attached to a player's character or location is metadata. Even with Magic the Gathering, a popular collectable card game, the way cards correspond to each other are like "events," and that material creates a giant database of information. Additionally, "to get real nerdy in here," Renee adores the Dewey Decimal System and has always been very organized. They volunteered at their local library from a young age and figured out the book classification system with minimal coaching. They feel structured thinking is the way their brain is structured to work. Everything since then has led to thinking in databases or tables, which they create in their adult life.

In the world of GIS, Renee underscores two essential skills: A willingness to experiment and the ability to accept feedback and criticism. "You have to realize that things are not going to come out right the first time. I enjoy being able to play with my work. Change colors, change fonts, customize things, that is the fun part of work for me." We also discussed the balance between objective data analysis vs. subjective art, a principle on which GIS is based, and how that sometimes comes with pushback. On receiving feedback, they shared, "Even if it's constructive, you still don't necessarily want to hear it. We aren't wired to want criticism... This is part art project, and I am going to miss something. Something will mean something [else] to another person."

Renee has been involved in a lot of projects throughout their career, but their favorite project was as part of FirstMap, Delaware's comprehensive self-service Enterprise Geographic Information System. They worked on creating a radiological emergency tool; in the event of an emergency nuclear situation, this tool allows the users to view a nuclear weapon's potential impact on concepts such as population and agriculture in the state of Delaware. Notably, some of their work on this tool extended to the federal level.

Renee is most passionate about the utilization of GIS for disasters and disaster recovery. They feel that there's a great potential for GIS to be used not only to track what areas are hit the hardest, but to also ensure that those affected are getting equal access to items and resources during an emergency. The use of unmanned aerial vehicles to capture data safely is how they believe GIS can coincide with emergency and disaster services and responses, as well as what the future of GIS looks like.

Renee has a long list of people to thank. A lot are GIS Departments that they have had the pleasure of working with or are current coworkers. Specifically, "[One mentor] was so awesome to have as an instructor, as someone who really believed in why and what [they] wanted to learn and did not wave [them] away for already having a graduate degree and returning to community college. She reassured [Renee] that learning was not linear."

They continue growing in GIS by taking training courses, some of which include programming languages Python and R. They stay up-to-date with GIS on LinkedIn and talk to GIS professionals at their job. Renee is hoping to continue their formal education with an additional Master's degree in either Disaster Management or General Planning.

To Renee, "GIS means Connection. It means a way to connect people onto a common playing field." Map literacy has improved substantially over the years, and GIS map applications can now be easily accessed and used by everyone. By providing data on accessibility, GIS also aids people with health conditions or impairments; Renee believes that "mapping what is accessible and what isn't is huge."

Their advice to all others starting in GIS is as follows:

"It's not just about you, it's about living a mile in those that you are serving's shoes".

ANGELICA JOYCE

Angelica has always had an interest in science and geography courses. She received Associate's and Bachelor's degrees in Geography with a minor in Climatology and Meteorology from Southern Illinois University Edwardsville. While pursuing her Associate's, she learned about GIS, her earliest understanding of which was that it was an interactive map with interactive layers. She feels GIS is a very employable skill by which she is deeply fascinated, saying she "took a passion and translated it into something that is employable."

One of Angelica's personal interests lies in geopolitical events; specifically, the Russian invasion of Ukraine. She "had the privilege of working on a project of mapping Ukraine at the beginning of the war, and it was amazing connecting [her] passion with [her] career and doing something impactful."

Throughout our conversation, Angelica consistently emphasized her desire to engage in meaningful work. Her passion for GIS is undeniable, and she seeks continuous growth while maintaining a balance between her professional pursuits and personal interests. Her perspective on GIS is clear: "I think you have to have a focus on learning and a want to learn. It's always changing and there is a lot you can do with it."

A project she wants to highlight is her capstone project for her Bachelor's in Geography. She compared Eastern and Western Germany and investigated topics such as income, migration trends, total wealth, and gender disparity across various divisions of the areas. Even now, "They still are different. It was a big project to undertake."

Her ideal goal would be to obtain security clearance and work with classified data in GIS. A "pipe dream" would be to work for any world government and live abroad. Historic GIS work fascinates her, but most importantly, she would like her work to incorporate her passion of making an impact in the world. She also sees a possible future in atomic energy and feels it is of growing importance for our future. On this path, she would like to help in the effort of constructing power plants and finding safe locations to store nuclear waste. For Angelica, undertaking these efforts "is one of the keys to future sustainability." Ultimately, she wants to work where the biggest and most impactful GIS work is located.

She would like to thank a small group of women - a group through which we personally met. In the male-driven STEM world, she feels having a support network is not only highly enjoyable, but essential to success. On the connections she has made in GIS, Angelica states that she's met "some really cool people in GIS," including "goobers and really cool people, living life and doing GIS work!"

Her advice to all others starting in GIS is as follows:

"Try to do anything and everything that you can. Learning in the workplace, while fun, is also stressful. The more well-rounded you can be the better".

JONATHAN MORTENSEN

Jonathan graduated from the University of St. Thomas with a Bachelor's of Science in GIS and a minor in biology and sustainability. He completed several projects while working towards his undergraduate degree, including land use change in cities over a 10-year period using aerial imagery, the path of foreign chemicals into an aquifer, and a conservation plan for a local natural conservation area. His capstone focused on limiting urban flooding in danger zones.

His early understanding of GIS was with online interactive maps and models of Earth, which can be used to calculate routes between you and your destination. His understanding broadened in college, as his professors used GIS regularly. He found that a career in GIS would expand the environmental and conservation work he would like to do.

Jonathan's personal interests align with GIS quite well. He is "a bit of a gamer... a very online person." He likes strategy games with overhead visualization; he says this allows him to maintain a GIS mentality while playing. In his free time, he uses programming languages like HTML and Python to code, both of which are used frequently in GIS.

A project he would like to highlight was his work on calculating the ideal locations for planting trees in an urban environment. Using critical thinking, he figured out what features he should use to get the best return of investment: Streets and roads, park trails, topography, temperature, fine particulate matter in the air. Jonathan "actually used a logarithmic scale with bike trails and roads [to see] how far away from those would be best for planting these new trees." He ran a comparison script through a model builder to check criteria for potential planting zones. Jonathan took a step further and used his passion for biology to decide which species of trees should be planted.

There are three skills he feels are essential to finding success in GIS: Critical thinking, creativity, and patience. He says that when a client presents you with an idea of their final product, you have to work backwards to discover your workflow and execute the client's vision. Sometimes, a unique request allows him to try something new. He feels you can express your personal creativity not just by picking colors on the map, but by finding a creative solution to the problem. When working with large databases, he feels users should be forgiven with the software, as there is a learning curve: "Maybe I am not an absolute saint, but everyone should have enough patience to survive in GIS."

Although Jonathan admits he comes from more of an environmental background, in college he became more interested in the human geography aspect of GIS. Combining the two, he would like to work towards sustainable urban planning. Overall, he wants to pursue an oversight role on an operation that would benefit everybody. His favorite aspect of GIS is "the macroscopic scale of data visualization you can work with. The biggest and best thing about it is how many different variables of real-world space you can compare and contrast." He has put significant time into learning database competencies and working with SQL to control and manage large sets of data.

He believes the future of GIS will be working with three-dimensional data and that Light Detection and Ranging (LiDAR) data will allow us to do 3D models of entire landscapes. He discussed the implications of a 3D map for hikers, where they would no longer be looking at a two-dimensional overhead view; instead, they would look at the landscape, hills, and features. He wants to learn more about Indoors GIS and visualizing 3D spaces.

He now believes, "GIS means macroscopic, total holistic view of the world's information. It is the world at your fingertips." You can scale the entire world to different projections; whatever data you put in, you can get something out.

Jonathan would like to thank his college professors who used GIS and have stayed in contact with him and other alumni through LinkedIn.

His advice to all others starting in GIS is as follows:

"Diversify your knowledge". He feels GIS is a very generalist field that can be used in any discipline for multiple reasons. "Learn some programming, learn database management, querying, get skills that allow you to work in any industry".

MADISON FLORY

Madison completed her undergraduate studies at Kansas State University, earning a Bachelor's degree in Geography. Her academic journey initially began with a focus on Physical Science, but she found herself drawn to Geography as she delved into elective courses, eventually making the switch in her senior year.

Reflecting on her early exposure to GIS, Madison recalls, "When I first heard the word GIS, I had no idea [what it was]." She initially likened it to GPS but soon realized its broader scope. Her initial reservations gave way to enthusiasm after enrolling in more advanced courses, leading her to conclude that GIS "is one of the best tools [she] could use to further [her] career and [her] passions."

Motivated by a desire to make a positive impact on a wide scale, Madison envisions her educational background as a way to benefit society at large. She is more interested in the practical impact of her work than personal recognition. Madison aspires to contribute to roles that support first responders or aid in disaster relief, ultimately aiming to save lives through her GIS expertise.

In addition to her GIS pursuits, Madison embraces diverse hobbies and interests that enrich her personal and professional growth, even if they aren't directly tied to GIS. She describes herself as a "jack of all trades" who enjoys reading, painting, photography, and hiking. These varied interests facilitate meaningful connections and expand her network, enabling her to explore a wide range of opportunities.

When it comes to the qualities essential for success in the GIS field, Madison emphasizes the importance of unwavering determination. She believes that one must be resolute in the face of software challenges, stating that "you have to be more stubborn than the program you are working with." Critical thinking is another vital trait, especially when navigating the complex problems that arise from GIS software.

Despite her limited work experience, Madison has been actively involved in GIS-related projects, both during her studies and early in her career. In collaboration with a small group, she created maps showcasing the overlap of five major water pollution types with demographic data. Additionally, Madison worked on mapping bird migration patterns, comparing contemporary and historical data to shed light on the impacts of climate change on bird behavior.

Madison's deepest passion lies in the realm of natural disasters and how GIS could contribute to disaster management. She envisions herself working on disaster relief and emergency management projects, driven by her desire to provide essential GIS data to support first responders like her father. Madison dedicates her time to volunteering for disaster management using an open-source GIS software called Open Street Map (OSM), where she helps map real-time data in disaster scenarios, enabling rapid response efforts.

Additionally influenced by the work of a geographer she follows online, Madison is inspired by projects involving tornado debris characteristics and trajectories. This work uses social media to track the scattering patterns of tornado debris, providing insights into the development and evolution of tornadoes during extended storm events.

Looking to the future of GIS, Madison believes that GIS will become integral to nearly every industry. While she anticipated the integration of AI assistance, she underscores the continued need for human expertise to effectively operate GIS software. She sees GIS as a field with immense potential for growth, expecting it to expand further as new and improved software solutions emerge.

To stay current in the field and foster her continuous learning journey, Madison plans to take advantage of free online GIS courses provided by various software platforms. She aims to acquire new skills and knowledge to become a more proficient GIS analyst. For Madison, GIS has evolved into an all-encompassing tool that has already made a positive impact on her life and has the potential to influence her future in unpredictable ways. She believes that persistence is key to reaping the rewards of completed GIS projects and to experiencing the satisfaction of making a tangible difference.

Her advice to all others starting in GIS is as follows:

"You've got to stick with it. It's very, very rewarding to see that final project that you spent the last X amount of time working on come together and become one cohesive thing. [You'll think] 'I did that' and it is a very, very rewarding feeling. [...] At the end of the day, the hard work and the frustration is 100% worth it."

ANDREW BROZ

Andrew is a graduate of the University of Florida, where he received a Bachelor's of Science in Forest Resources and Conservation, primarily specializing in environmental resources, wildlife, and forestry. His notable exposure to GIS took place over a single intense semester, during which he took a GIS class complemented by an associated GIS lab. Throughout this course, he created increasingly intricate maps, which culminated in a project designing a hiking trail traversing several African countries, meticulously highlighting natural areas of hiking interest.

During his initial exposure, Andrew didn't have a deep understanding of GIS. His focus was predominantly directed towards following a traditional career path focused on biology fieldwork. However, the mandatory GIS course opened his eyes to its potential applications across various fields. He encountered a thought-provoking experiment from one of his professors, who challenged students to identify a field where GIS held no relevance, only for the professor to skillfully demonstrate how mapping could indeed be beneficial. Andrew was captivated by the versatile applicability of GIS across diverse fields. He began to appreciate the value of GIS in the field, considering it instrumental for analysts in enabling them to comprehend the intricacies of environmental phenomena.

Outside the realm of GIS, Andrew's personal and professional growth stems primarily from his extensive 40-year career in forestry, which significantly enriched his understanding of the natural world. His profound connection to the outdoors, cultivated through hiking and exploration, played a pivotal role in shaping his passion for GIS. In his leisure time, he plans sustainable land management strategies for private landowners, converting his forestry career into a hobby while he continues to advance his GIS career. Some of his forestry maps can be quite complex, sharing, "I kind of go overboard a bit but I really enjoy it."

Essential qualities and character traits for success in GIS, according to Andrew, include logical analysis to catch potential errors, the ability to swiftly discern patterns, understanding future applications in the real world, and empathizing with end-users to deliver effective solutions. These attributes underpin his work, ensuring precision and practicality, particularly when dealing with intricate networks like roadways.

Andrew's GIS-related projects predominantly revolve around forestry, spanning "cover type maps, inventory, delineating timber sale boundaries, and tracking wildlife populations." His maps, crucial components of state-certified legal contracts, provide clarity, simplicity, and visual accessibility for end-users.

His passion for GIS centers on creating seamless, precise, and visually appealing representations of features, paying great attention to details like watersheds and tree line curves. Andrew's career goals revolve around leveraging GIS for environmental uses, with aspirations of becoming a reforestation analyst. He says, "I just want my work to be used for environmental betterment." He pursues knowledge through workplace experience and is eager to absorb insights and skills from his professional environment.

To stay updated with GIS developments, Andrew relies on professional training and opportunities. For him, GIS is a means of contributing to environmental well-being without becoming overly disheartened by the state of the ecosystem. In his opinion, LiDAR technology represents one of the most exciting possibilities in GIS. Its applications in wilderness studies and forest management fascinate him, particularly with its utilization of baseline data and expansion into remote, inaccessible mountain ranges.

His advice to all others starting in GIS is as follows:

"Take every opportunity that you can. If it's something outside of your comfort zone, jump on it. Don't just stick to the same analysis that you're doing. [...] Make yourself invaluable."

IAN STAUFFER

Ian holds a Bachelor's degree in Geography with a concentration in GIS and a minor in Geology, obtained from Central Michigan University. During his undergraduate journey, Ian delved into advanced graduate-level courses and embarked on an independent study project within the realm of geology, which involved the mapping of air pollutants.

His initial exposure to GIS came from a basic geography project during his first year at Central Michigan. As he engaged in fundamental mapping, his interest was piqued by the possibilities of expanding the horizons of cartography. Seeking guidance, Ian approached his professor, who directed him towards the university's GIS courses.

Ian's motivations for pursuing a career in GIS crystallized when, a year or two later, he enrolled in a course titled "Geomathematics." In this course, he used mathematical techniques to map pollutants in water data, specifically focusing on tracking their dispersion within the groundwater table over time. The ability to visualize the progression of toxic materials in groundwater and convey this critical information to relevant authorities, such as the EPA, underscored the transformative potential of GIS.

While at Central Michigan, Ian collaborated with the city planning board of Mount Pleasant. Their project involved an in-depth analysis of affordable housing and zoning in the city, crucial inputs for rewriting the city's master plan. Ian's specific role included mapping the population density of Mount Pleasant and overlaying it with zoning maps, shedding light on zoning appropriateness concerning high-density populations and low-income housing.

Outside the GIS sphere, Ian's personal and professional growth is nurtured by his involvement with the Godot game engine program, a passion that enhances his programming prowess, particularly in languages like JavaScript and Python. Additionally, his competitive engagement with Pokémon, both in video games and trading cards, has sharpened his analytical skills. Ian and a friend created a tournament-specific spreadsheet to dissect usage rates, moves, and strategies, demonstrating his dedication to rigorous analysis.

In the context of GIS, Ian deems qualities such as a need to understand everything presented, an inclination to dig deeper into problems, and a knack for spotting connections as indispensable. These traits align with spatial awareness and spatial statistics, enabling him to excel in the field. He shares, "I don't like 'alright well that's a Monday problem', I just don't like leaving 'well enough' alone; I like digging deeper [and] finding out the actual connections to things."

Ian's passion in GIS revolves around the development of remote sensing, code-intensive aspects and the automation of complex processes, with a focus on the intricate facets of coding. To enhance his skills, he remains current with Python updates and audits geospatial analysis courses.

His career aspirations span environmental GIS and planetary sciences, with a particular penchant for extraterrestrial mapping in realms like asteroid and moon cartography. "I'm making sure I keep my hand on the pulse of the industry," particularly advances in LiDAR technology. Ian anticipates exciting prospects in GIS, particularly in machine learning and AI applications. While acknowledging current ethical considerations, he foresees the acceleration of image classification and remote sensing data analysis, allowing GIS professionals to explore other facets of their work.

Notable influences in Ian's GIS journey include Carl Sagan, an icon in the field of astrophysics, as well as a passionate professor from Central Michigan University who actively engaged students in his ongoing projects. This hands-on approach left a lasting impact.

Ian's commitment to continuous learning includes reading professional newsletters and tracking companies involved in off-planet exploration. To him, GIS provides the ability to discern correlations among seemingly disparate elements and recognize their interconnectedness.

His advice to all others starting in GIS is as follows:

"Keep an eye out at various companies that do what you're interested in and see what they hire for [...]. See what they want from people and try to keep your extracurriculars in line with that."

WODY GALLO

Wody graduated with a Bachelor's of Science in Applied Geology from MSU Denver. Despite starting college without a defined path, she drew inspiration from her father, a remarkable physicist/chemist whose research at Mass General Hospital in Boston aimed to combat cancer. Her passion for applying knowledge to make a positive global impact was heightened after her father's passing. Consequently, Wody chose Geology as her major and began her adventure with GIS while completing various projects, including those related to Geophysics and lithology maps. She quickly developed a deep affection for crafting maps and visual representations to complement her projects.

Her early encounters with GIS were somewhat limited in understanding, as she mainly associated "it with maps and geography but never connect[ing] how important the understanding of GIS is towards benefitting the world." However, as her education progressed, she recognized the profound significance GIS has in shaping the world. This realization that GIS beautifully combines technology with the power of map-based comprehension fueled her determination to pursue a career in GIS.

Her geospatial sciences professor at MSU Denver profoundly influenced her understanding of GIS. His engaging online courses, enjoyable projects, and informative video lectures illuminated the incredible potential of GIS as a career. An assignment to research GIS jobs online opened her eyes to the myriad of possibilities within the field.

Despite her limited work experience, Wody was actively engaged in GIS-related projects during her studies. These endeavors encompassed diverse topics such as geophysics, remediation projects for dredging, mapping volcanic hazards, topology, and population data. A particularly memorable project involved identifying the optimal location for building a Hogwarts school, which required an adherence to strict parameters and secrecy from the "muggle world."

Beyond her GIS journey, Wody is an avid outdoors enthusiast, finding solace in hiking and nature exploration. These activities allow her to employ maps to navigate diverse terrains, fostering her connections to both geology and GIS.

For Wody, meticulous attention to detail is a cornerstone of GIS success. She thrives in an independent environment, relishing the opportunity to focus intently on tasks without distractions. She feels this is a characteristic well-suited to GIS, which often offers remote work options.

Wody's passion for GIS centers on its environmental applications, particularly in contributing to remediation and conservation efforts. However, she realizes that "there are an insurmountable number of possibilities in this career field that can better the world." Her undergraduate projects, including a Computer Aided Design (CAD) cell report project focused on dredging pollutants from Salem Harbor, significantly enriched her expertise in this area.

Wody believes that advancements in aerial imagery and real-time data represent exciting trends in GIS. Real-time data in particular holds immense potential for detecting and responding to small changes across various scientific domains, from position and temperature to concentration and pressure.

To remain aware of the latest developments in GIS, Wody

subscribes to numerous GIS newsletters. Currently, Wody is seeking employment opportunities to further hone her GIS skills, with aspirations of pursuing graduate studies in the field. To her, GIS is not just a career; it is a way to use her love for science to positively contribute to the world.

Her advice to all others starting in GIS is as follows:

"Though the work can be challenging and detail oriented it is so rewarding knowing you're doing something to improve the lives of even one person. If you love science and computer technology this is the perfect career for you. It is necessary and combines all the sciences while using advancements in the technological world. Some of the programs can be frustrating at times, but it is always worth it when producing a product you can be proud of".

GARRETT CARR

Garrett Carr hails from Utica, OH, and earned a Bachelor's of Science in GIS from Ohio State University. However, Garrett's journey into GIS wasn't initially apparent. He entered the field with little understanding of its nuances, coming from a background in engineering. His pursuit of an engineering degree gave rise to challenges in coding and mathematics, leading him to seek alternative avenues that were both "technical and had long-term growth potential in the workforce." The hyper-competitive nature of the engineering college at Ohio State only added to the complexity.

During his undergraduate experience, a class about cartography stood out as a pivotal experience. Through this course, Garrett had the opportunity to discover the world of geospatial and satellite software, including notable commercial and open-source tools like QGIS. Garrett's affinity for the latest GIS software soon became evident, as the maps he crafted possessed not only aesthetic beauty, but also exceptional clarity and interpretability. ERDAS, with its use of infrared and satellite imagery, presented a unique dimension that also interested him.

Progressing in his GIS journey, Garrett delved deeper into the subject through "Geovisualization." This course equipped him with the skills to analyze and explain intricate mapping details, enabling him to formulate effective strategies for creating professional, accurate, and concise maps. Subsequently, "Introductory Spatial Data Analysis" allowed him to expand his horizons by the creation of spatial data charts, tables, pie charts, bar graphs, and more. Mathematical formulas became his tools in effectively articulating and comprehending spatial data.

Garrett found his calling in GIS, confidently stating that GIS will always be needed and "has an umbrella to include so many different career paths within it." Garrett's professional evolution reflects this diversity, as he transitioned from working with satellite imagery to engaging in survey utility fieldwork, both of which prominently feature GIS applications. His perspective emphasizes the importance of long-term thinking in career decisions, a mindset cultivated through personal experiences of financial struggle during his childhood.

During Garrett's formative years, he played baseball, basketball, soccer, and golf. These activities taught him a fundamental lesson: "There's no 'I' in team." Moreover, Garrett's sports experiences provided him with invaluable leadership insights. He realized that assuming a leadership role carries a weighty responsibility—one that extends beyond what most anticipate. He additionally honed essential interpersonal and communication skills, recognizing their universal significance in both personal and professional spheres. Garrett firmly believes in the golden rule of treating others with respect and kindness, which "will usually always be reciprocated."

Garrett shares that compassion, genuineness, empathy, understanding, professionalism, respect, and adaptability are all qualities he both personally values and he feels underpin success in GIS. He emphasizes the importance of remaining true to oneself, advocating for authenticity as a cornerstone of professional conduct. For Garrett, companies should value individuals for who they are, rather than expecting conformity to a perceived ideal. These principles are not confined to the workplace, however; he also advocates for their practice in daily life, where they may eventually become second nature. Garrett approaches his work with unwavering commitment, treating each day's tasks as if they were his last—an approach rooted in his belief in living life to the fullest, free from regrets and filled with dedication.

Though Garrett's work experience may be relatively limited, he has been involved in GIS-related projects. He undertook contract work involving satellite imagery for the Department of Defense. He is also actively using LiDAR to scan utility poles and collect information for use by his current employer's engineering department. Although he is currently working on the data collection side of GIS, he would love to transition into creating final GIS products.

Garrett's passion for GIS knows no bounds; he finds inspiration in every facet of this diverse field. With the inherently expansive nature of GIS, Garrett remains open to exploring the vast possibilities it presents. In his inaugural year within the workforce, he remains in a state of flux, uncertain about which specific path to pursue. He aspires to broaden his horizons while continually enhancing his skills and knowledge base. Garrett believes that his ongoing development will render him a versatile, sought-after candidate in the job market, setting the stage for a fulfilling career. He approaches each day as an opportunity to gain wisdom, eagerly absorbing insights from both colleagues and everyday experiences.

His early journey in GIS has been profoundly shaped by the guidance and wisdom of experienced professionals. His gratitude extends to former colleagues from his first position held in the GIS industry. Their invaluable mentorship and insights have left a lasting mark on his understanding of GIS.

Garrett envisions a promising future for GIS, particularly as he embarks on his new role in the utility sector. He recognizes the enduring importance of utilities as the essential infrastructure of the world, which fuels his excitement about the GIS industry's ongoing growth and vitality. Garrett believes that the world's reliance on various utilities ensures a sustained demand for GIS applications, underscoring the profound impact GIS will continue to have in shaping and supporting the essential services upon which society depends.

He plans on staying updated on the latest developments in GIS by networking on LinkedIn. To Garrett, GIS is "the lifeline of the world." He appreciates how he can use it to impact people directly and indirectly, ultimately making a true difference. Garrett "can go to sleep every day knowing that [he] made a difference and that is good enough for [him]."

His advice to all others starting in GIS is as follows:

"Be YOU and stay true to who you are. There will be times where you think all is lost and question plenty of decisions in your life. Never let it steer you away from your passion and what you love. Never allow yourself to be judged by how many times you get knocked down but by how many times you get back up."

CONCLUSION

Summary of Common Themes:

Throughout this comprehensive exploration, we've identified several recurring themes that unite the journeys of these ten rising GIS professionals:

Diverse Pathways: The paths to GIS are diverse and unique, ranging from geology to environmental science, from biology to geography. These varied backgrounds enrich the field, highlighting GIS's interdisciplinary nature.

Passion and Purpose: These rising stars' collective narratives reinforce the importance of passion, purpose, and the desire to make a positive impact on the world through GIS. Their dedication to creating meaningful work is palpable.

Skills and Learning: Those featured consistently emphasize the need for continuous learning and skill development in GIS. Adaptability and a commitment to expanding one's knowledge base are core to success.

The Evolving GIS Landscape: The GIS landscape is ever-evolving, and these individuals exemplify the adaptability required to thrive in this dynamic field. They embrace emerging technologies including LiDAR, real-time data, and aerial imagery, recognizing their transformative potential.

Community and Networking: Building a supportive network within the GIS community is crucial, whether through professional organizations, social media, or mentorship. Collaboration and shared knowledge are essential to teamwork and collaboration.

A Call to Action:

These comprehensive narratives may serve as a profound source of inspiration for individuals embarking on their GIS journeys, reinforcing the importance of curiosity, resilience, and the pursuit of meaningful work. GIS is not merely a career, but a means to make a tangible and positive difference in the world.

Acknowledgment of Contributions:

I would like to express my heartfelt gratitude to all featured rising GIS stars: Will Campbell, Renee Hupp, Angelica Joyce, Jonathan Mortensen, Madison Flory, Andrew Broz, Ian Stauffer, Wody Gallo, and Garrett Carr. Their generosity in sharing their stories, insights, and wisdom has illuminated the diverse and dynamic nature of the GIS field and will hopefully inspire and educate others.

Final Thoughts:

As we conclude this extensive exploration of the rising stars in GIS, we are reminded that GIS is a field of boundless possibilities, where technology, science, and creativity converge to address real-world challenges. It serves as a bridge between data and the real world, offering innovative solutions to complex problems. The future of GIS is filled with optimism and excitement, shaped by the unwavering commitment of those featured. Their collective vision underscores the limitless potential of GIS to contribute to a better world.

Advice for Aspiring GIS Professionals:

- Follow your passion and align it with your GIS journey.

- Embrace continuous learning and skill development.

- Build a supportive network within the GIS community.

- Stay open to new opportunities and remain adaptable.

- Find meaning and purpose in your GIS work.

ABOUT THE AUTHOR

Georgia Mackay is a self-proclaimed "GIS Miracle Worker".
She lives in Middletown, Delaware with her two cats Pepper and Bean, her
bearded dragon Pancake, and her boyfriend John. In her free time, she
enjoys helping others, playing airsoft, and solving sudoku puzzles.